THE SUNSTONE SUPERSTOVE

A Safer, More Efficient Way of Heating With Wood

by John Lago

Loompanics Unlimited
Port Townsend, Washington

Neither the author nor the publisher assumes any responsibility for the use or misuse of information contained in this book. It is sold for informational purposes only. Be Warned!

The Sunstone Superstove
A Safer, More Efficient Way of Heating With Wood
© 2003 by John Lago

Cover photo by John Lago
Interior illustrations by Jim Blanchard

Published by:
Loompanics Unlimited
PO Box 1197
Port Townsend, WA 98368
Loompanics Unlimited is a division of Loompanics Enterprises, Inc.
Phone: 360-385-2230
E-mail: service@loompanics.com
Web site: www.loompanics.com

ISBN 1-55950-237-1
Library of Congress Card Catalog Number 2003106880

Contents

Introduction

When fuel costs started to escalate during the seventies the realization that you can't really be comfortably warm unless you're also economically comfortably followed. Over the past few years the high cost of operating gas, oil, and electric heating systems has taken the comfort out of keeping warm and replaced it with a feeling of dread and guilt. The higher the setting on the thermostat, the more one got the shivers when thinking about the utility bill.

For many, a switch to wood was the answer. Wood was cheap or even free for the cutting in many areas, plus it gave the user that blessed feeling of independence. No doubt independence and economy are the main reasons that over twenty percent of the houses in North America are today equipped with some sort of a woodburning system, be it a backup or an only source of heat.

As attractive as wood-burning may seem in the beginning, it soon becomes apparent to anyone who has ever used wood as a main or only source of heat that there are a number of drawbacks connected with this type of fuel.

At the top of the list is the danger. Even a properly installed stove has the potential to quickly turn your house into a smoking pile of charcoal and ashes. Locate combustibles too close to the unit, for instance, or fail to clean the flue on a regular basis, and it's likely that sooner or later you'll be standing beside your smoking ruins mouthing the equivalent of "I didn't know it was loaded."

Ironically, sometimes this danger increases with experience. After a time, one gets so used to thinking of ye old woodburner as a friend that one begins to "trust" it. After all, it rescued you from The Big Brother Oil and Gas Company, right? And didn't it heat the water for your hot chocolate during a day of fun and games in the snow? And what about the romance of a crackling fire? You come in tired and numb on a cold day, throw on a few logs and sit back and enjoy. Oh what a feeling. What a deadly feeling. Forget the Alamo. Remember the Trojan Horse? It is alive, 1400°, and beguiling you with deadly trust. (Note: All temperatures mentioned herein are in Fahrenheit.)

Inconvenience is another drawback of the ordinary woodburner. Even if a source of wood is readily available, it still has to be cut, split, dried, and finally carried inside and burned on a regular basis before it can do any good. The catchword here is "regular." Regular burning can get to be an insufferable drag, especially near the end of a demanding winter. If a crackling fire is romantic, then this part is the headache. "Not tonight, Honey," you want to call out to the stove when the in-house temperature is rapidly falling at 4 a.m., "I've got an ice cream headache."

The difficulty in maintaining a stable temperature is another drawback associated with woodburners. A common complaint is the house is either too hot or too cold. The freeze-fry effect. If it's not a shriveling 88°, then it's so cold that all the liquids

Introduction

in the house are in danger of turning into solids. Different kinds of wood put out different amounts of heat. Regulating the draft and damper can be tricky. Maintaining an even, livable temperature overnight with most woodburners is not an art and is not a science. It is an accident.

Part of holding a steady temperature depends on the regulation of the damper, which leads us to another drawback of the ordinary woodburning system: air pollution. Fill the firepot and close down the draft and damper on any woodburning stove, and what you have is a fair amount of heat radiating into the house and an unfair amount of particulate matter spewing from the chimney. Not only is this particulate matter mucking up the chimney and the atmosphere and causing breathing difficulties in communities where much wood is burned, but from a monetary standpoint you're getting less heating value per cord of wood. Sometimes fifty percent less. This is called the political way of burning wood. Lots of waste. Very taxing. Yet choking the fire is necessary to maintain a stable temperature, you say. A necessary evil. Evil, maybe, but not so necessary, as you will soon see.

So. If wood burning has been around as a source of heat since prehistory, and is still around, why can't a system be built that eliminates the well-known major drawbacks of burning wood? Why doesn't someone, with or without the technology available, put together a simple system that eliminates most of the danger, the inconvenience, and the pollution? Why doesn't someone do all this, plus (and this is a big plus) build into this same simple system a thermostatic control so that exact temperatures can be *dialed* instead of guessed at? Why doesn't someone do this?

Well, why don't *you* do it? What you're reading right now is designed to help you realize this goal. In simple Dick-and-Jane terms and paint-by-number illustrations these pages will provide you with the information necessary to build a Sunstone Superstove.

Not only that, but these pages will provide you with the information to do it inexpensively. Since one of the major concepts of the Sunstone Superstove is economy, it follows that building it should not cost an arm and a leg. In fact, most of the material used to put this thing together will be material that other people throw away. Junk. Scrap. You can even call it garbage if you like, but in the end it will become a thing of function, beauty, and simplicity, and with a small amount of maintenance it will stay that way far beyond the number of winters you care to dwell on earth.

Figure 1 is a general, simplified illustration showing the workings of a Sunstone system. As you can see at a glance, there is nothing really technical or mysterious about the function or design. You may be in doubt about some of the details involved in actual construction and operation, but more about that later.

For now, think of the Sunstone Superstove as working like this:

1. You heat a volume of stones with a wood fire.

2. The fire chamber and the stones are separate, but both are contained within an insulated shell.

3. When your house needs heat, a thermostat activates a fan, which blows heat out of the stone battery and into your living spaces.

4. When the desired temperature is reached, the thermostat stops the fan until the house again needs heat.

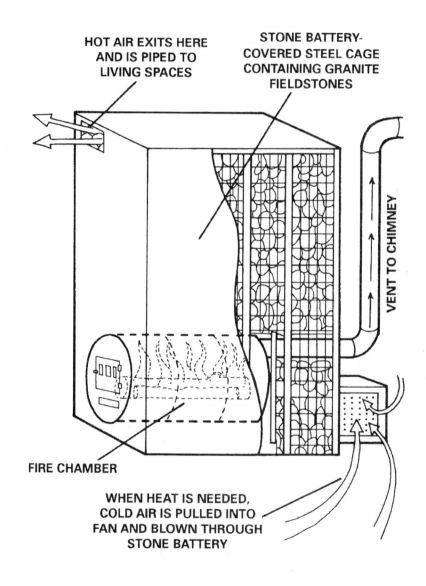

Figure 1

Explanation of Features

For the sake of clarity and credibility, exactly how will this unit reduce the danger, the inconvenience, the pollution, increase the efficiency and maintain a stable in-house temperature all at the same time? It sounds like a pretty tough assignment for such a simple arrangement of materials, but let's go over these pertinent points one at a time.

The danger. Like a naughty child, a woodburner causes most of its mischief when unsupervised. Look at the statistics. The most damaging fires occur when the operator is:

1. Not at home.

2. Asleep.

3. Drunk.

While we won't elaborate on the third reason for the incidence of damaging fires, consider these two common scenarios:

1. You leave the house with a fire still burning in the stove. While you're away, a spark is carried to the chimney where it lodges in an accumulation of creosote and starts a chimney fire. Since nobody's around to snuff the flames in the early stages, they gather intensity, crack the chimney, and spread to the surrounding structure.

2. It's a bitterly cold night. Before you go to bed, you overload the firebox in rightful anticipation of continuing cold. An act of foresight has you placing an armload of kindling and other firewood near the stove since you know the first thing you're going to do when you wake up in the morning is start another fire. Unfortunately you don't wake up in the morning because sometime during

the night your overheated woodburner ignited the wood set out for the next fire and you died in your sleep of smoke inhalation.

The list could go on and on — and it does every year in the newspapers — but the point is leaving a fire unattended increases the likelihood of disaster in excess of 90%.

A Sunstone system eliminates most of this danger by allowing you to plan and attend your fires according to your schedule and not according to the dictates of the weather. For instance, if you're home during the evening hours, and at work or asleep the rest of the time, then you do a burn during the evening. To create enough heat to last the night and into the next day, you burn all the wood you'd ordinarily save for the night and the following day during these evening hours. Chuck it in and burn it up. The excess heat will be absorbed by the stone battery.

Of course, when a whole day's supply of wood is burned in, say, a four-to-eight-hour period, this will naturally result in a much hotter fire. Which brings us to the claim of *lesser air pollution* and *greater efficiency*.

Pollution from a wood fire is, quite simply, smoke. Smoke, in turn, is essentially the airborne byproducts of combustion, consisting mostly of carbon in the form of minute particles of soot and droplets of tar. When wood is heated, "smoke" is given off. If sufficient air is supplied to the smoke, and the smoke itself is hot enough (about 1100°), it combines with oxygen in the air, or "burns," and heat is given off. The key words here are "heat" and "air." A hot fire supplied with plenty of air is the cleanest way to burn, since most of the smoke is being burned in the stove instead of being vented unburned up the chimney.

The benefits of a hot fire don't stop there. At the same time a hot fire is decreasing pollution, it is also *increasing efficiency*. Roughly half the heating value in a given amount of wood is contained in the "smoke." That is, if the "smoke" is driven out of the wood and vented up the chimney unburned — as in a smoldering fire — not only are you cluttering up the community air space and your chimney with soot and tar, but you are also losing out on up to half of the wood's potential to heat your house.

To be sure, any stove can be burned "hot." It's only a matter of open drafting and venting. Burn most stoves hot for any length of time, though, and soon the kids' crayons are melting and a glass of water only has to be heated about twelve more degrees before it boils.

A Sunstone furnace, on the other hand, is *designed* to burn hot. "Excess heat," or heat not immediately needed to warm the living spaces is not released. It is stored in the stones.

Much inconvenience and worry is eliminated with this method of burning and storage. No longer is it necessary to anticipate the weather or select a certain size and type of log according to the length of time spent away or asleep. You do a hot burn with whatever wood is available at a time when it is convenient for you.

As far as *maintaining a stable in-house temperature*, this is the job of the stones, aided by the insulation and the thermostat and fan. Stones have long been known to have a huge capacity to absorb and retain heat. In fact, a pound of granite can hold more than three times the amount of heat than can be held by a pound of steel. Put to use in the proper way, stones can serve as a storage for heat much as a battery can serve as a storage for electricity. When generating electricity, excess juice can be stored in a battery. When generating heat, this excess can be stored in a volume of stones. If heat from a woodstove is radiated into a volume of stones, these stones naturally rise in temperature. And if this temperature is effectively contained, it can be piped off on a regulated basis with a fan controlled by a thermostat. In other words, complete control of room temperature, regardless of whether the wood fire is burning almost red hot or not burning at all.

Specifics

As I said, Figure 1 is a general illustration showing the basic design and operation of a Sunstone system.

For a fire chamber it has the familiar 55-gallon drum, or "barrel stove," as it is sometimes called. Although almost any front-loading and rear-venting wood stove of moderate to large capacity can be used to charge the stone battery, a 55-gallon drum has many built-in advantages.

First of all, these drums are common to the point where even the best of them sometimes qualify as trash. So anybody looking for one doesn't have to look far, and when he finds one the price is usually right.

Another advantage of the standard drum is its proximity to the floor when installed horizontally as a stove. (Since the stone battery can only be charged with the stove *below* the stones, a stove with a low profile is a must.)

A sound barrel stove installed and operated with a little common sense will last a surprisingly long time. Keep it dry and burn only wood, and the unit can be expected to last seven years or more.

Ready-made leg, door, and vent kits are available in most hardware or discount stores.

A barrel stove will accept logs up to 30″ in length — no small consideration when you stop to think how this feature saves on needless cutting and splitting.

As mentioned earlier, the stone battery is where the heat is stored. Not all types of stones are suitable for use here, and only a few are ideal. Sedimentary rocks, such as limestone, are not recommended since their capacity to store heat is low, and they may have internal cavities that could explode when heated to the temperatures of a fully charged stone battery.

Without getting into a technical treatise on the drawbacks of other types of rock, let's just say that igneous rocks — rocks formed under great heat and pressure — are the only way to go.

These include diamonds, rubies, quartz, marble, and granite — with granite being at the top of the list. Granite is hard, dense, resists dusting and flaking, and is odorless when heated. It is also very common. So common, in fact, that in many regions (the Canadian Shield, parts of the Midwest and the Northeast) it is a genuine curse. The same size and shape of granite rocks (3″-9″ rounded) that are ideal for use in a stone battery litter some parts of the landscape in such profusion that you can't walk three feet in a straight line without tripping on one. Since rocks are hell on plows, cows, and low-slung sows, most landowners are more than pleased to have you come in and haul away as many as you need, and then some.

And how many do you need? That depends on how much heat you want to store.

First of all, we'll lay down the ground rules on heat and stones.

Heat is measured in BTUs. "BTU" stands for British thermal unit. Measuring heat in BTUs is the accepted standard.

One pound of granite will hold .38 BTUs for every degree Fahrenheit its temperature is raised.

One cubic foot of granite stones weighs about 115 pounds.

Heat that cubic foot of granite to 230° above room temperature (300°) and it will store a little more than 10,000 BTUs.

(.38 BTUs x 115 lbs. x 230° = 10,051 BTUs.)

Notice in the calculations that we used *230° above room temperature* instead of the figure 300°. That is because once the granite cools to 70° (room temperature) it will no longer give off any "usable" heat. Heat only flows from warmer to colder, so once the granite drops to 70° (or 230° below 300°), the heat flow stops.

We can conclude then, that one cubic foot of granite stones heated to 300° (maximum recommended temperature of the stone battery) will store 10,000 usable BTUs. Every added cubic foot will store another 10,000 BTUs. So you can see *it's basically the size of the stone battery that determines the amount of heat you can store.*

Now that you know that every cubic foot of granite in the stone battery will store 10,000 usable BTUs, the next order of business is to find out how much heat you *want* to store. How much heat you want to store depends on two factors:

1. The heat demand of your house.

2. How long you want to maintain a warm house between fires.

An easy and accurate way to determine the heat demand of your house (how many BTUs it's using) is to request an energy audit. They're free, and they're performed by your local utility company. By inspecting your existing furnace, and by comparing the efficiency of that furnace with the figures on your utility bill, the people at the utility company can give you a fairly accurate estimate of the heat demand of your house. They can tell you how many BTUs are being used on a typically severe winter day, a mild day, or almost any day in between. Call them. It's part of their job.

The information provided by an energy audit, and the knowledge that every cubic foot of granite is worth 10,000 BTUs, is your starting point in building a Sunstone Superstove.

The size is your choice.

Do you want to build a unit with a capacity to store heat for several days? A week? Or would you be satisfied with something designed to hold only enough heat to get you warmly and safely through the danger period between bedtime and breakfast?

Once again, the size is your choice, but before we get into the actual nuts-and-bolts information concerning the construction of a Sunstone system, let's talk about the average house.

The Average House

You may balk at the mention of an "average house," but think of the type of dwelling that has dominated the housing market since the 1950s. No doubt you will picture a three-bedroom ranch, bi-level, or whatever. It will have from 1100-1500 square feet of floor space, 6 inches of insulation in the attic and 3½ in the sidewalls. It will also have double pane windows and a basement. Your average house.

About seventy percent of the heated houses in the U.S. and Canada fall under this heading. Average. Maybe not average in looks, configuration, or decoration, but average in the sense of a similar heat requirement.

Introduction

In the northern tier of states and in Canada, this typical house can be kept comfortably warm from day to day in severe winter weather by building a Sunstone system supplied with two cubic yards of rocks. Heat this volume of rocks to 300° on a sub-zero Monday evening and it's very doubtful that the members of the household will have to suffer room temperatures less than 70° until sundown Tuesday when you'll have another fire. In other words, 16-18 hours of stored heat.

Naturally, during milder weather, two cubic yards of rocks heated to 300° will satisfy a heat demand for a much longer period — sometimes for several days.

You might say then, considering all the average houses out there in BTU-land, and considering that many potential burners-of-wood would like to sail cozily through the night and into the next day without having to poke, feed or fret over a fire, that a heavy favorite for many would be the two cubic yard unit.

For this reason, detailed plans for building a two cubic yard unit will be featured in the following pages. It will have a storage capacity of about 500,000 BTUs.

As you no doubt have guessed by now, the weight of one of these heaters will not allow it to be installed just anywhere. To quote one benefactor who built a five cubic yard unit, "You might as well put it in the basement, because if you put it in the middle of the kitchen floor, it's going to end up in the basement anyway."

Even though it will be exerting fewer pounds per square inch on a floor than a masonry chimney, it'll still need much more support than most ordinary woodburners.

Chapter One
The Two Cubic Yard Unit

A completed unit built according to these plans will occupy a floor space 68" square and stand 75" in height.

Location. Locate on a clean, bare concrete floor. If the floor has vinyl tiles, carpeting, linoleum, etc., pull this covering up and remove any glue or fillers still adhering to the concrete. Besides being a fire hazard, any floor covering or adhesive left on will later cause odors when heated.

It's recommended that wood-burning appliances be placed no closer than three feet from any combustible surface. Combustibles not only include wood, paneling and the like, but also gypsum wallboard, plaster and lath, and any plastic or fiberglass resin material. Check clearances. This means *heat-radiating surfaces*. On a Sunstone furnace the exposed heat-radiating surfaces are the door and the uninsulated surfaces surrounding the door, and the uncovered flue pipe between the rear of the unit and the chimney. The flue pipe requires an 18" clearance between it and a combustible surface.

The rest of the exterior of the finished unit will not radiate heat and does not fall under the above clearance requirements.

Try to locate the unit about six feet from a chimney rated to handle wood fires. Why six feet? Since a properly operated furnace will be burning hot, the stack pipe between the furnace and the chimney will also be hot. It's a fact that from 20-30% of a fire's heat is radiated from the first six feet of stack pipe. If you can keep this stack pipe 18 inches from combustible surfaces on its way to the chimney, plenty of heat that would ordinarily escape up the chimney will instead be radiated into your living spaces.

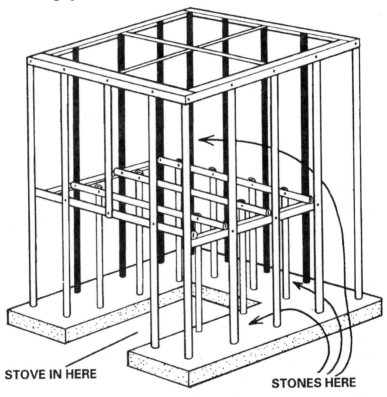

STOVE IN HERE

STONES HERE

Figure 2

The first order of business is to build a framework for the unit. You'll need angle iron, steel pipe, and an assortment of nuts and bolts. Finished, the framework will look something like Figure 2.

What Figure 2 shows is two sturdy cages made of pipes with the bottom ends set in a bed of concrete. The stove slides inside the smaller cage. The rocks are placed in the space between the smaller cage and the taller, larger cage.

This arrangement allows for easy, efficient absorption of heat from the stove to the stones.

The framework is made from angle iron, steel pipe, and flat steel stock. (Figure 3.)

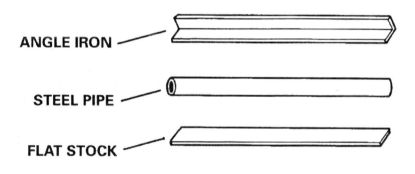

ANGLE IRON

STEEL PIPE

FLAT STOCK

Figure 3

The top of the framework is fashioned from angle iron and flat stock.

Needed:

4 angle irons, 2″ x 48″

2 pcs. flat stock, $^1/_8$″ x 1″ x 47″

The outside framework is made from steel pipe.

Needed:

15 steel pipes, 1" x 63"
1 steel pipe, 1" x 35"

The inside framework is made from steel pipe and angle iron.

Needed:

8 steel pipes, 1" x 29"
6 steel pipes, 1" x 45"
2 angle irons, 2" x 36"

A word here about angle irons and steel pipes: Bought new, this material can be very expensive. To hold costs down, use scrap. Scrap yards usually have used pipe and angle iron in quantity, and always at a very slight fraction of the cost of new material. Like about ten percent. It doesn't matter if the available material is black iron, galvanized or slightly rusted. Avoid heavily rusted or pitted iron. Avoid too, thin-wall electrical conduit pipe. It lacks the necessary strength.

The easiest way to assemble the stone cage is by starting from the top.

Take the material needed for the top — the four 2" x 48" angle irons and the two pieces of flat stock, and lay them out on a flat, level surface as shown in Figure 4.

Weld at all joints.

(Note: the rest of the framework is put together with nuts and bolts, but for the sake of rigidity and continued squareness in assembly, the top should have welded joints. If you don't have a welder or have a friend who does, a welding shop can do the job in a few minutes for a few dollars.)

Make sure the job is square. A square layout will have equal diagonal measurements from corner to corner.

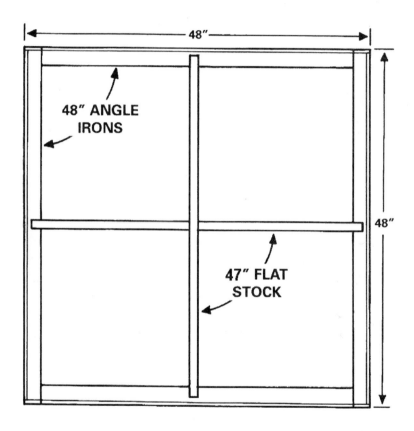

Figure 4

You now have a square, welded top piece. Before you can bolt up the 15 vertical pipes that will form the outside framework, you'll need to drill and countersink 15 holes in the angle iron. The holes should be ¼".

Looking at the top piece, from the ends, drill and counter-sink holes as shown in Figure 5.

Countersink so that a bevel-headed ¼″ bolt inserted in the hole is flush with the surface of the angle iron.

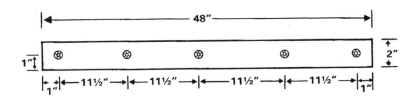

(SIDES) 5 HOLES ¼″

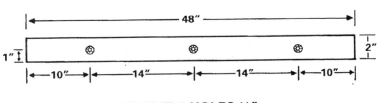

(FRONT) 3 HOLES ¼″

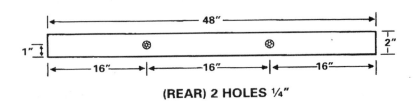

(REAR) 2 HOLES ¼″

Figure 5

Remember when I said that the easiest way to assemble the rock cage is by starting from the top? True. But to do this you first have to suspend the recently welded and drilled top piece from the ceiling.

Suspend it over your choice of location about 62″ off the floor — just high enough so that when you bolt on your 63″ pipes, the bottom ends are dangling about a half-inch off the floor.

To suspend the top piece, all you need is two 5-foot 2 x 4s, four shorter pieces of wood, and a few nails.

Hang it as shown in Figure 6, making sure it is level from side to side, and from front to back.

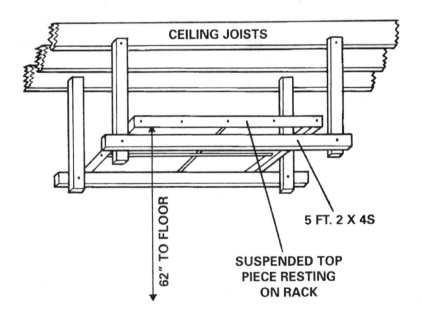

Figure 6

Now you're ready to bolt on the pipes that will form the outer cage.

Drill a ¼″ hole through the fourteen 1″ x 63″ pipes, ½″ from one end. Do this also with the 1″ x 35″ pipe. (See Figure 7.)

Figure 7

Using a bevel-headed ¼″ x 2″ bolt, fasten the 35″ pipe to the *middle hole* of the *front* side of the suspended top piece.

Using the same kind of bolts, fasten the fourteen 1″ x 63″ pipes to the remaining fourteen holes drilled into the top piece.

Fasten the pipes *loosely*. Do not tighten yet.

The pipes are fitted against the *inside* of the angle iron as shown in Figure 8.

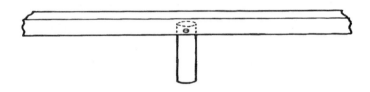

Figure 8

As shown in Figure 9, these pipes when hanging perpendicular should not be touching the floor. Shim up the top piece if they do happen to be touching.

FLOOR **ABOUT ½" SPACE**

Figure 9

Before the bottom ends of the pipes can be set in concrete, they must first be held firmly and squarely in place. For this you'll need to make a simple jig from 2 x 4s with the dimensions given in Figure 10, being sure that the *inside* of the jig measures 48" on a side.

Note: To facilitate the building of this jig and the forms on the following pages, the lengths of the individual pieces of formwood are given in the diagram. This is assuming that you are using 2 x 4s with an actual measurement of 1½" x 3½".

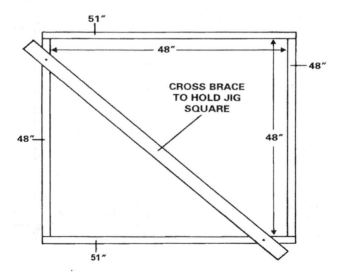

Figure 10

Fit the jig around the pipes.

Support the jig about a foot off the floor.

Clamp the corner pipes to the inside corners of the jig. Draw the clamps just tight enough so that you can tap the framework into alignment if the pipes are not plumb or if the framework is not quite square.

When you are satisfied that the corner pipes are plumb and that the framework is square, tighten the corner clamps and wire the remaining 63″ pipes to the jig. (See Figure 11.)

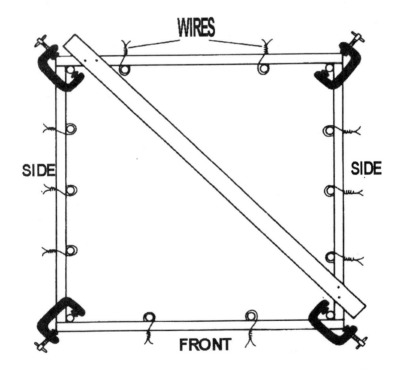

Figure 11

With the jig in place (about a foot off the floor) and the pipes comprising the outer framework clamped, wired, and square, turn your attention to building the forms for the concrete that will hold the cage permanently in place.

Using 2 x 4s set on edge, form an area around the pipes as shown in Figure 12. The forms will be 3½" deep.

(In this illustration, the pipes are shown *as they are held by the jig.* For the sake of clarity in showing the dimensions and position of the form, the drawing of the jig itself is left out.)

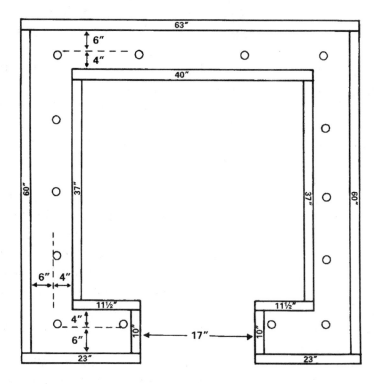

Figure 12

When you're satisfied that the form is in its proper place, you might want to set some concrete blocks or some other heavy objects at different points along the outside edges to keep it from moving while you're pouring the concrete.

Now for the concrete. For this pour (there will be another smaller pour later for the inside framework), you'll need 3.8 cubic feet of concrete.

If you live close to a batch plant where ready-mixed concrete is available, go there with six empty five-gallon buckets. They'll fill them for you, cheap, and just like that you will have your required 3.8 cubic feet of concrete. All that remains is to drive the buckets home and fill and strike off the forms.

Other than that, you can mix the concrete yourself in a wheelbarrow, either with a bagged pre-mix, or by putting together your own sand-gravel-cement mix.

Pre-mixed bagged concrete is usually sold in 40 or 80 lb. bags. All the ingredients are in the bag except the water. An 80 lb. bag makes $^2/_3$ cubic foot of concrete. Therefore for this pour at least six 80 lb. bags will be needed. Expect it to cost about 20 dollars.

A cheaper way to go is to mix the concrete from scratch. Pick up a 94 lb. bag of Portland cement, six five-gallon buckets of ¾ inch gravel, and four five-gallon buckets of sand.

To mix, dump a five-gallon bucket of gravel into a wheelbarrow, about 3½ gallons of sand, and about a gallon and a half of cement powder. Add about a gallon of clean water, gradually, and mix thoroughly with a garden hoe. Don't get it too wet. Stop adding water when the mix is a little on the wet side of crumbly.

Shovel it into the form and mix another batch. When the form is full, tap the edges with a hammer to settle the concrete and to drive out the air pockets.

Strike off the top with a board or trowel, making sure the concrete is level with the top of the form.

After letting the concrete stand for about three or four days, strip off the forms, the jig, and the material used to suspend the top piece.

This completes the outside framework.

Now that the outside framework is complete, let's go on to assembling the inside framework.

Take the two 2″ x 36″ angle irons and bolt them to the outside cage 28″ from the floor, as shown in Figure 13. Use ¼″ x 2″ carriage bolts.

You'll have to clamp the angle irons to the pipes at the premeasured height and drill through both pieces of stock while they are clamped together in order to get the holes to line up.

(Before you drill, make sure the horizontal top of the angle iron is 28″ from the *floor*. Not from the concrete you just poured, but from the floor itself. See the circle in Figure 13.)

Using 2 x 4s, build a form as shown in Figure 13. The outside dimensions of this form will measure 20″ x 34½″. The area between the outside of this form and the inside edge of the previous pour will be filled with concrete once the rest of the inside framework is assembled.

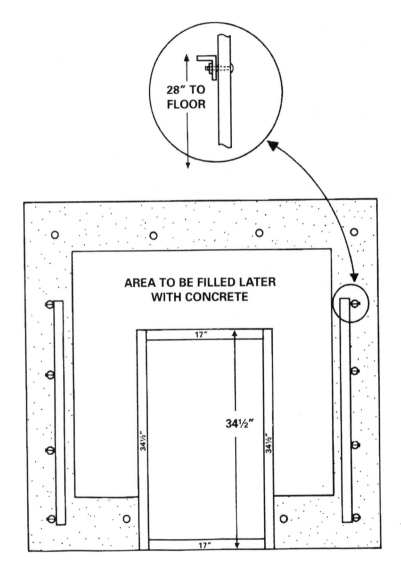

Figure 13

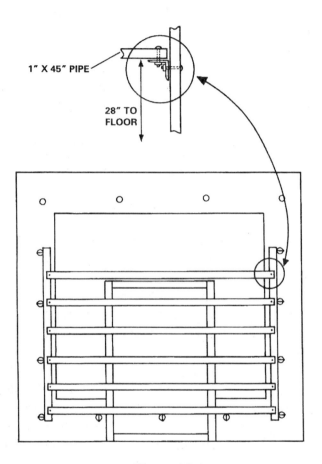

Figure 14

In Figure 14, the six 1″ x 45″ steel pipes are resting on and spanning the distance between the angle irons you just bolted up.

Clamp, drill, and bolt these pipes to the angle irons as shown, using ¼″ x 2″ carriage bolts.

(See Figure 2 for an illustration showing a different perspective of the same layout.)

When in place, these pipes will be about 6″ apart, on center, but it's not necessary that they be exactly 6″ apart.

Now take the eight 1″ x 29″ pipes and drill one ¼″ hole in each pipe, ¾″ from an end, as shown in Figure 15.

¼″ HOLE ¾″ FROM END

Figure 15

These eight pipes are to be attached to the 1″ x 45″ pipes that were used to span the distance between angle irons.

Mark and drill the 1″ x 45″ pipes so that the 1″ x 29″ pipes can hang vertically as shown in Figure 16.

(Tight quarters may require that you temporarily remove three of the 1″ x 45″ pipes after they are marked, to correctly drill the holes.)

Using ¼″ x 3½″ carriage bolts, attach the 1″ x 29″ pipes. Hanging vertically, their bottom ends should be dangling about a half-inch from the floor. (See detail of Figure 16.)

You're now ready to fill the area between the outside of the wooden form and the inside of the previous pour. This will take about 2.3 cubic feet of concrete, or a little more than half the amount needed for the first pour.

After the pour, put a short level on the eight 1″ x 29″ pipes extending vertically down into the fresh concrete and make sure they are *reasonably plumb*. You won't need a jig to hold these pipes exactly in place as you did with the outside cage, since they won't be holding any further building material that would require them to be square and plumb.

After letting the concrete set for a few days, knock the forms away and tighten all bolts.

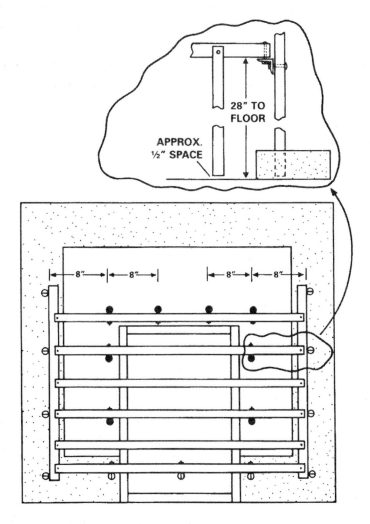

28" TO FLOOR

APPROX. ½" SPACE

8" **8"** **8"** **8"**

Figure 16

The Firebox

The first thing to do is to find a sound, standard-sized 55-gallon drum. Standard-sized drums are 35″ long and measure 23″ in diameter.

When you've found what you think is an acceptable drum, it'll need six additions before you can position it within your recently completed stone cage.

1. A fire door.

2. An ash door.

3. A grate.

4. A fitting for the flue pipe. (Called a thimble.)

5. Legs.

6. An extension pipe. (Explained later.)

As mentioned earlier, ready-made leg, door, and vent kits are available in hardware or discount stores. They are usually made of cast iron and are a once-in-a-lifetime purchase. That is, when a barrel eventually wears out, you simply transfer the kit parts to a new barrel and you're back in business.

There are different types of kits on the market. When choosing, keep these three points in mind:

1. The thimble. You need a *rear*-mounting thimble — not a top mounting one. It has to bolt up to the *rear* of the barrel, and it should be the type that the stack pipe fits into, and not around.

2. The ash door must be able to *swing open*, so as to allow the operator to reach in with a shovel and scoop out the ashes that fall below the grate. (However, an "ash door" that doesn't meet this requirement can be

easily modified with a simple hinge and latch if a true ash door is not available.)

3. The legs included in most kits will hold the barrel too high for use in a Sunstone setup. If the legs in your kit can't be cut down to hold the bottom of the stove no more than 2″ off the floor, you'll have to fashion your own legs from four pieces of flat steel stock. (This procedure will be explained below.)

Grates are ordinarily not included in barrel stove kits, since most operators of these stoves place a bed of sand or firebrick on the bottom and build a fire on top of this.

For the sake of a quick, clean, and complete burn, though, having a grate in a Sunstone furnace is *very important*. Very important and very easy to construct. Details on construction will be given later. (See Figure 17.)

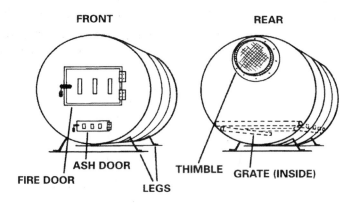

Figure 17

Let's say you've found an acceptable barrel stove kit. Before making the cuts on the front of the barrel to accommodate the fire door and the ash door, keep in mind that the grate, when installed, will rest 4½″ above the bottom of the stove.

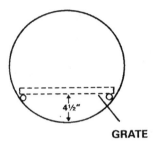

Figure 18

This means the cut-out for the ash door has to be made below the grate line, and the cut-out for the fire door has to be made above the grate line. (See Figure 18.)

If the legs included with your barrel stove kit can't be shortened to hold the stove no more than 2″ off the floor, a set of serviceable legs can be made with four pieces of flat steel. (See Figure 19.)

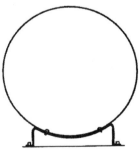

Figure 19

To arrive at this configuration, take the two 26″ pieces and bend and drill as shown in Figure 20.

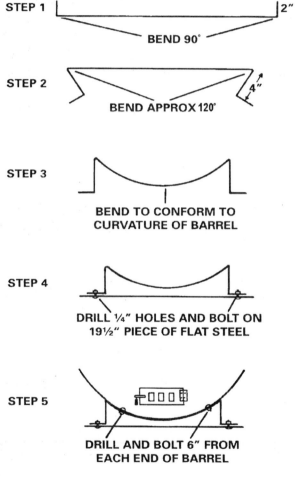

Figure 20

The Grate

Having a grate in a stove gives an operator an extra degree of control over a fire. A grate suspends the burning wood, allowing air to feed and fan the fire from beneath. Quick, clean, and steady fires can be built and maintained with the air fanning from below the grate. Almost 100% combustion efficiency can be achieved once the fire is up to an operating temperature of about 1100°.

To build a suitable grate for a barrel stove, you need:

(2) ¾″ pipes, 32½″ long

(17) ¾″ pipes, 17″ long

(34) ¼″ carriage bolts, 2½″ long

Drill $^5/_{16}$″ holes in the 32½″ pipes, starting 2″ from one end and continuing at 2″ intervals, 17 holes in each pipe. (See Figure 21.)

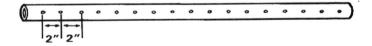

Figure 21

Drill two $^5/_{16}$″ holes in the 17″ pipes, ½″ from each end. (See Figure 22.)

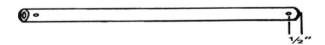

Figure 22

Since the grate, when assembled, is 17″ wide and the opening cut for the fire door will measure less than 17″, the grate might have to be at least partly assembled inside the stove — a clumsy procedure at best.

To save a lot of cussing and fumbling, first try bolting the grate together loosely outside the stove, then twist it as shown in Figure 23.

Twisted thusly, it may fit through the fire door where it can then be straightened and set in place. Bolts need only be finger tight.

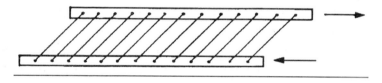

Figure 23

In position, the grate should look as shown in Figure 24:

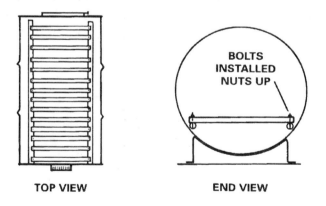

TOP VIEW   END VIEW

Figure 24

When the fire door and ash door, the thimble, and the grate and legs are installed to your satisfaction, the next step is to attach a 32" section of extension pipe to the thimble.

Note: a word here about the extension pipe. This particular piece of pipe is the connecting link between the stove and the stack pipe that will later lead to the chimney. It, like the stove, will be radiating heat into the stone battery once the stove is operational. Since this pipe is not going to be visible once the stove is operational, and since it is by function and design considered to be a working extension of the stove itself, it should be heavier than your ordinary garden variety of stack pipe. Twenty-gauge, or thicker, is recommended. A sheet metal shop may be the best source for this pipe.

Mate the extension pipe with the thimble. Angle the pipe upwards so that the open end is about 1" higher than the thimble end.

Secure the joint with three sturdy sheet metal screws. It's best to drill through the pipe and thimble while they're tight together, staggering the holes at equal distance around the circumference so that the extension pipe can't be moved up, down, or sideways once the screws are turned in. (See Figure 25.)

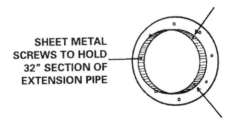

SHEET METAL SCREWS TO HOLD 32" SECTION OF EXTENSION PIPE

Figure 25

Important: This unit must be "burned off" when completed. Carry the stove outside and build a hot fire on the grate. See that the fire spreads the entire length and width of the firebox. Many small pieces of wood work best for this job, rather than a few large pieces.

Keep the fire hot by adding more wood until all paint, oil film, and any other combustible coating has flaked or burned off the exterior of the barrel. Brush off any remaining dust or residue when the unit cools down.

Carry it back inside and slide it into the recessed area of the rock cage. Push it back until the rear of the barrel butts up against the concrete. (Tip of arrow in Figure 26.)

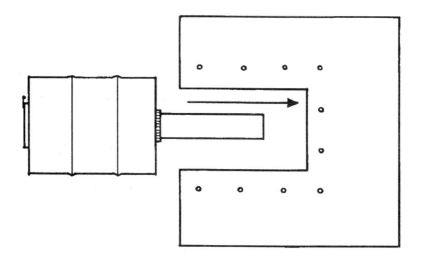

Figure 26

Lining the Stone Cage

Since the pipes comprising the stone cage are placed too far apart to hold in the size stones you'll be using in the stone battery (3"-9"), the cage will need a wire mesh liner. Twelve-gauge wire with a 2" x 4" rectangular weave is recommended. (See Figure 27.)

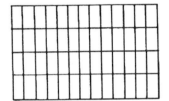

12 GAUGE WIRE
2 X 4" WEAVE

Figure 27

The easiest way to go about this is to buy a 5 x 25 foot roll, lay it out flat and cut 10 pieces according to the dimensions given in Figures 30A and 30B.

Tip: Since you're going to join the ends of these individual pieces inside the stone cage, you might want to cut the mesh so that — on certain ends — 2" spines of extra wire are extending beyond the given dimensions. (See Figure 28.)

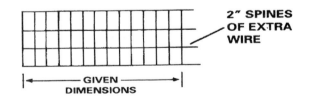

2" SPINES
OF EXTRA
WIRE

GIVEN
DIMENSIONS

Figure 28

These spines extending from one piece of mesh form a good splice when twisted around an adjoining piece of mesh. (See Figure 29.)

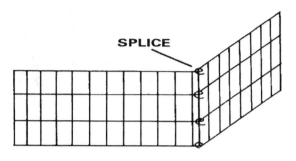

Figure 29

So, in all but one of the illustrations in Figures 30A and 30B, the end *or ends* where the 2″ spines should be extending are marked by showing the extensions in heavier ink.

When cutting the mesh, just remember that the spines are not included as part of the given dimensions. The spines, where shown in heavier ink, will add about 2″ of extra length to the given dimension.

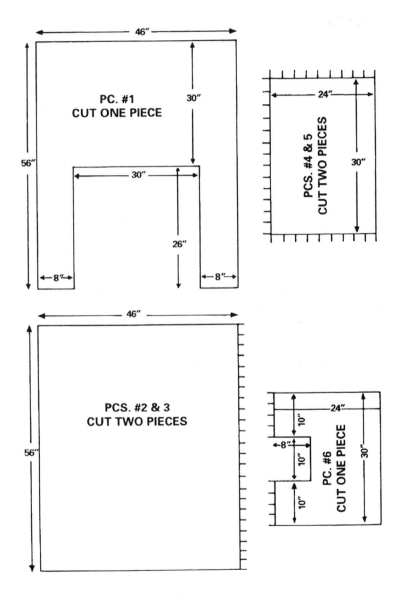

Figure 30A

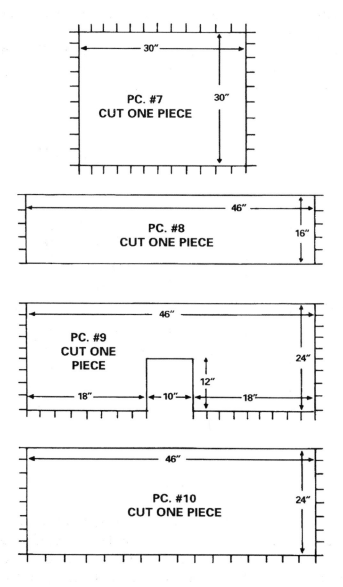

Figure 30B

You'll notice that each of the 10 pieces of mesh are assigned a number from 1 to 10.

Starting with piece #1 and ending with piece #7, line the stone cage with the mesh as shown in Figure 31, splicing the ends at 4″ intervals to make strong seams.

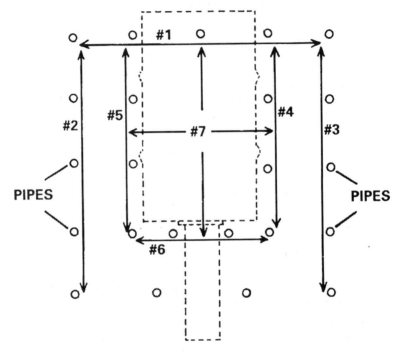

Figure 31

Pieces #8, 9, and 10 will be put in later. They cover the backside, but for the time being the backside should be open, because the next step is to add about half the stones. It has been found that these stones are much easier to lay in place if one can simply walk through the back of the stone cage to place them rather than having to crawl over the top.

Prospecting for Stones

Suppose you don't already have in mind a good source for stones. Let's take a few minutes here to elaborate on one of the easiest known methods of procurement.

As mentioned earlier, 3"-9" rounded granite fieldstones are ideal. "Hardheads," they are often called.

During the last ice age the glaciers scraped countless millions of these stones up from the earth and scattered them over half of the U.S. and Canada. Farmers have been skimming them off their fields for centuries. Sometimes they've taken the trouble to use the stones as fences between fields, but most of the time they simply dump them at the end of a field to get them out of the way.

This is where you come in. Take a little drive out in the country on a spotting expedition. If you live in the Northeast, the Midwest, or in eastern or middle Canada, chances are you'll barely be beyond the city limits before the stone piles pop into view. Find one close to the road and ask the landowner if he has any designs on it. If he didn't get a copy of this book before you did, he probably has no designs on it other than to find a way to make it disappear.

So you both make out. He gets rid of a bunch of granite garbage, and you come home with a pile of nuggets that are like gold itself when translated into the warmth and comfort and safety they'll provide over the course of a lifetime.

Filling the Stone Cage

Clean the stones thoroughly. A hard spray from a garden hose is usually enough for most moss and dirt, but a little work with a stiff brush might be necessary when it comes to clinging clay and lichens.

A load of stones dumped at random will have about 40% voids. Put another way, if you fill your stone cage with no regard as to how the stones are placed, almost half the volume of the cage will be air. You don't want that. Stone density is what you want — not air space. The denser the arrangement of your stones, the more heat the cage can hold. It's all a matter of placement. For rounded stones, a good mix is the key. All small stones together will have about 40% voids no matter how you place them. The same with all larger stones. But the right mixture of smaller and larger stones can easily cut this naturally occurring void in half. It's only a matter of putting down several of the larger stones and then filling the spaces between them with smaller ones. Plug the voids. Taking a little time here will translate into longer times between fires later.

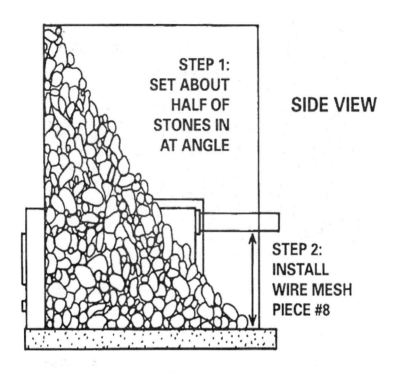

Figure 32A

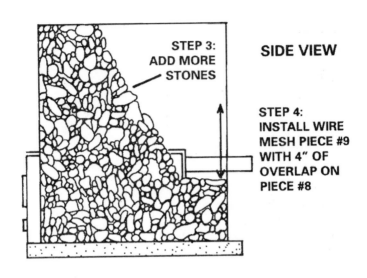

STEP 3:
ADD MORE
STONES

SIDE VIEW

STEP 4:
INSTALL WIRE
MESH PIECE #9
WITH 4" OF
OVERLAP ON
PIECE #8

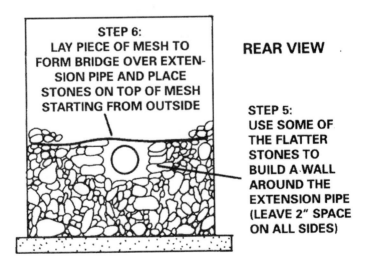

STEP 6:
LAY PIECE OF MESH TO
FORM BRIDGE OVER EXTEN-
SION PIPE AND PLACE
STONES ON TOP OF MESH
STARTING FROM OUTSIDE

REAR VIEW

STEP 5:
USE SOME OF
THE FLATTER
STONES TO
BUILD A WALL
AROUND THE
EXTENSION PIPE
(LEAVE 2" SPACE
ON ALL SIDES)

Figure 32B

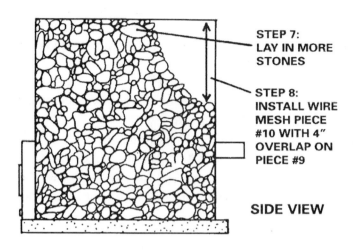

**STEP 7:
LAY IN MORE
STONES**

**STEP 8:
INSTALL WIRE
MESH PIECE
#10 WITH 4″
OVERLAP ON
PIECE #9**

SIDE VIEW

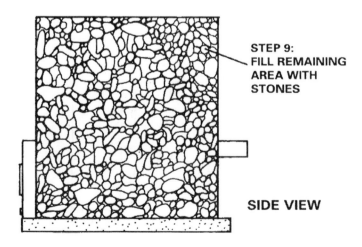

**STEP 9:
FILL REMAINING
AREA WITH
STONES**

SIDE VIEW

Figure 32C

Now that the cage is filled, you are looking at a stone battery capable of storing over a half million BTUs. With an insulating cover and a few controls you can almost look forward to the next cold spell.

First the cover.

Start by acquiring five pieces of galvanized sheet metal, one for the top and four for the sides.

Top: one sheet 48″ x 48″, 28- or 30-gauge.

Sides: four sheets 48″ x 60″, 28- or 30-gauge.

Set the top piece in place as it is. No need to use fasteners.

For the front, cut one of the 48″ x 60″ sheets as shown in Figure 33:

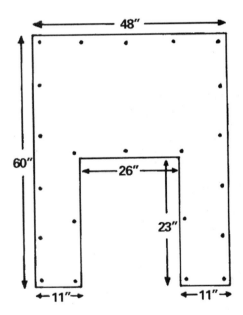

Figure 33

Stand it up against the front of the cage. At the four outside corners, drill through the sheet metal and into the pipes so that you can fasten the sheet to the pipes with short sheet metal screws.

With the sheet metal held in place at the four corners, drill through the sheet metal and the pipes at about one foot intervals around the edges. (Black dots in Figure 33.)

Turn sheet metal screws into the holes drilled.

Before one of the remaining 48″ x 60″ sheets can be fastened to the back of the stone cage, a cutout has to be made for the extension pipe.

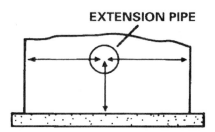

Figure 34A

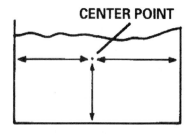

Figure 34B

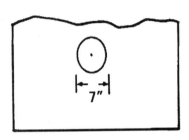

Figure 34C

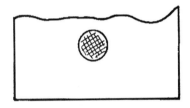

Figure 34D

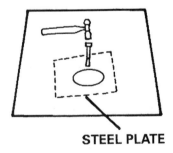

STEEL PLATE

Figure 34E

By measuring from the base and from both sides, locate the center point of the extension pipe. (See Figure 34A.)

Mark this point on the 48″ x 60″ sheet. (See Figure 34B.)

With a compass or divider, inscribe a 7″ circle around the center point. (See Figure 34C.)

Cut out circle. (See Figure 34D.)

Tip: Sometimes a small circle is easier to cut out with a cold chisel than with tin snips. To cut out with a chisel, lay the sheet out flat, with the area to be cut resting on a piece of steel plate. Using a hammer and a sharp chisel, tap around the line to be cut until the circle falls out. (See Figure 34E.)

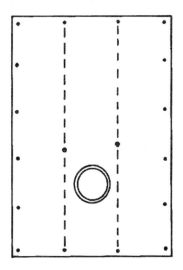

Figure 35

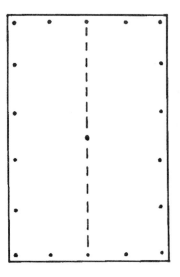

Figure 36

As was done with the front sheet, lean the back sheet up against the back of the stone cage and drill and fasten at the corners.

Then drill and fasten around the outside edge at about 12″ intervals. (Black dots in Figure 35.)

Also drill and fasten at the two points shown on either side of the extension pipe.

Using the same procedure, fasten the side sheets. (See Figure 36.)

A footnote here about sheet metal: 28- to 30-gauge sheet metal is recommended as a cover because it is thin — therefore lower in cost and easier to cut. However, almost any thickness of sheet metal will do.

For instance, if a scrap yard has large sheets which might be of a different thickness and considerably less expensive than new pre-cut sheets, there's no harm in putting them to use.

Also, there's no harm in using many smaller pieces instead of the abovementioned five large sheets, although using smaller pieces will cost you in construction time.

The Hot Air Outlet

Determine where you want to direct the heat flow. If you have an existing hot air furnace, chances are you've located the Sunstone furnace nearby and are planning to direct hot air from the stone battery into the hot air chamber of the existing furnace.

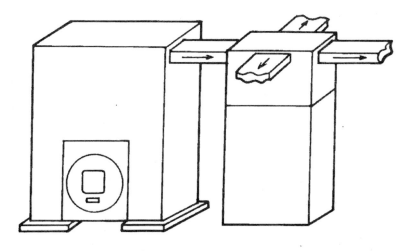

Figure 37

Or maybe you're building in your garage or workshop and all you want to do is feed hot air into a single room.

Regardless of where you plan to direct the heat, to be effective the hot air exit has to be near the top of the stone battery (since hot air rises) and it has to be fabricated so as to only allow heat to flow when needed.

A very simple and effective hot air exit can be made from an air grill, a small rectangular sheet of aluminum, and two small hinges.

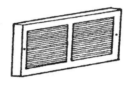

Figure 38

Air grills are sold in home building centers and look as shown in Figure 38. The kind you want is a plain one with no shut-off mechanism. It should have at least a 6″ x 12″ opening for the flow of air.

For about 70 cents, a sheet metal shop can supply you with the aluminum sheet. You want a flat, rectangular piece, 26- or 28-gauge, and large enough to cover the flow area of the air grill.

Hinge the aluminum sheet to the top of the air grill with two small hinges. (See Figure 39.)

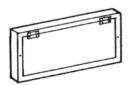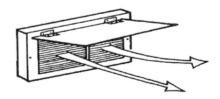

Figure 39

As you can see, the hot air exit operates by air pressure. When heat is called for, the fan (to be located on the lower opposite side of the stone battery) trips on and blows open the hinged aluminum flap and circulates hot air.

This modified air grill can be bolted up almost anywhere along the top of the stone battery — front, sides, or back — depending on where you want to direct the hot air.

Make the cutout at your choice of location. Be sure that the cutout is at least as big as the flow area of the air grill, 6″ x 12″, 7″ x 14″, etc.

Position the air grill over the cutout and fasten with sheet metal screws.

If you intend to connect to the hot air chamber on your existing furnace as shown in Figure 37, you'll have to make a cutout on the hot air chamber.

Then you'll need a section of ductwork to connect the two cutouts. Scrap yards are usually overflowing with different sized pieces of ductwork. Whatever your source, make sure the piece you choose is larger than the air grill. It has to fit *around* the grill so as not to interfere with the operation of the hinged flap.

The Cold Air Inlet

To create a good cross flow through the stone battery, the cold air inlet should be located opposite from the hot air outlet. For instance, if the hot air outlet is on the top left, the cold air inlet should be on the lower right. (See Figure 1.)

Before making the cutout for the cold air inlet at the bottom of the stone battery, it's best to first acquire a fan, since the fan discharge will hook up to the cold air inlet and you'll want these two openings to match in size and shape.

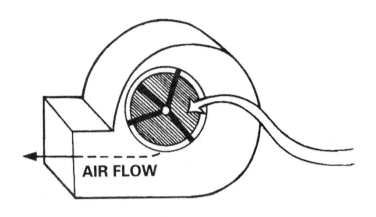

AIR FLOW

Figure 40

A drum, or "squirrel cage" fan from a regular forced-air furnace is ideal. This type of fan is quiet and efficient and looks like Figure 40.

Scrap metal yards are a good place to begin the search for a furnace fan. Find a complete, junked furnace and you'll likely find not only the fan but the drive motor too. Believe it or not, most of these motors and fans will be in good operating condition since furnaces are usually junked because of a burned out fire bonnet.

Be thorough in your search. New, a fan and motor can cost over a hundred dollars. But these two items used can sometimes be picked up for less than ten.

Notice that the discharge side of the fan is equipped with a "snout" of about 3″ in length.

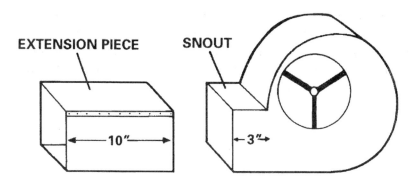

Figure 41

This snout will have to be extended about ten more inches to reach the cold air inlet through the layer of insulation and brickwork that will come later. (See Figure 41.)

Measure around the snout and cut and bend a thin piece of sheet metal to conform to the shape of the snout. Secure with sheet metal screws.

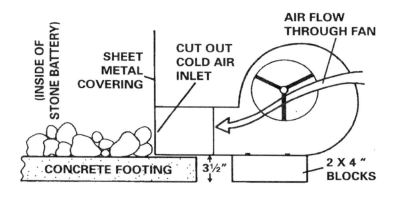

Figure 42

Attach two 2″ x 4″ blocks to the base of the fan to bring it level with the top of the concrete footing. Slide the whole works up against the sheet metal covering the stone battery. Mark around the fan discharge and cut a cold air inlet in the sheet metal. (See Figure 42.)

Chapter Two
The Exterior

In keeping with the tradition of building this furnace with things of value that others throw away, start thinking of a place where you can scrounge up some bricks. Regular building bricks. Bricks from a chimney someone tore down, bricks from a dump, or from a building being razed. Any kind of building bricks of the 2¼″ x 4″ x 8″ variety will do. You need about 600. They can be face brick, or common brick, or any combination thereof. Actually, for this job, the more varied the kind and color, the better the final effect.

And what effect are you after? Well, for one thing, these bricks are going to form the exterior of your furnace, so besides the durability and non-combustibility factor, you'll want to consider the visual effect.

In a word, it will be striking. You will be proud of it, and like the job itself, your pride will be lasting. The furnace as it stands right now probably suffers from a common case of the uglies, but you're going to change all that. With bricks picked up here and there, cleaned and set in place, you're going to finish off one of the safest and most convenient woodburning furnaces on the block and make it into a work of arresting visual impact. Others will shake their heads in skeptical admiration and say, "You did that?" and you will say, "Yup. Nothing to it, really. I just rigged up this simple little jig, collected these different bricks and then laid them down one on top of the other like here in this little book." Then you'll point to this section to show them what you mean and once again say, "Nothing to it. It's all in the setup."

The Setup

Brickwork would be simple if you didn't have to worry about keeping the job level and plumb. If all you had to do was mix a batch of mortar, lay it out in a line and set in the bricks, even the man in the rowboat could do it.

So, if you've never worked with brick, the following method will give you the advantage of not having to worry about keeping the job level and plumb. What you will do, essentially, is plumb and level the job ahead of time. You will make sure that all the rows will be straight and all the corners will be plumb before you mix a single spoonful of mortar or hoist your first brick.

To accomplish this, acquire six straight, wooden 1 x 2s (commonly called "furring strips" at home building centers), and 6 pieces of ½″ plywood, 4″ square.

Nail the furring strips to the plywood pieces as shown in Figure 43.

Figure 43

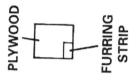

Figure 44

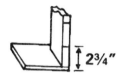

Figure 45

From the top they will look like Figure 44.

Starting from the bottom of the piece of plywood, measure and mark at $2^3/_4''$ intervals on one of the furring strips. (See Figure 45.) Go up to the 77″ mark. There will be 28 marks.

When you have one marked off, lay it down even with three of the others and mark them to correspond with the marks on the first one.

The remaining two strips don't have to be marked.

Look at Figure 48.

It shows how the six furring strips are set up to guide the brickwork.

It shows the *four marked strips* placed at the outside corners where the bricks meet, and the two *unmarked strips* butting against the bricks forming the front opening.

If you're in doubt as to the exact position of the furring strips, laying some bricks on the floor as shown in Figure 48 will give you a clearer idea.

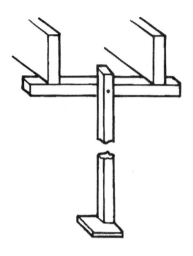

Figure 46

To keep the strips firmly in place while you lay the brick, they will have to be stuck to the floor and attached at the ceiling joists.

Squeeze a glob of construction glue under the plywood base plate. Set the base plate on the floor in its predetermined place. Attach the top of the strip to the ceiling joists as shown in Figure 46, using a level to make sure the strip is plumb.

When the glue dries, take a pair of pliers and push ½″ wire brads into the furring strips at the 28 marks made earlier. Push in two brads at each mark, as shown in Figure 47. These brads will hold your string lines.

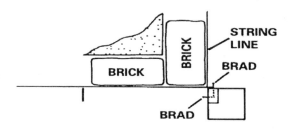

Figure 47

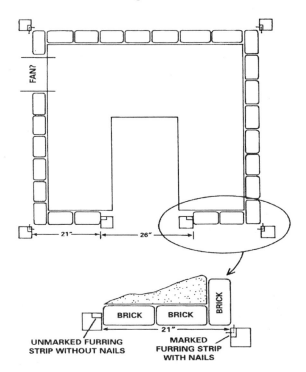

Figure 48

Now that the furring strip guideposts are set up, you can see that the work of keeping the brick-laying job level and plumb is basically taken care of. The guideposts themselves serve to keep the ends and corners plumb, and the string lines that will stretch between the guideposts will ensure that the rows can be laid straight.

Here's a tip on string lines: Use 6-8 lb. test nylon monofilament fishing line. There are two reasons for this.

1. Nylon monofilament is not as likely to collect small bits of stray mortar as, say, a multi-fiber cotton line.

2. Elasticity. Lightweight monofilament line can be stretched between two points to make a taut guide without exerting much pull on its two points of attachment — in this case the small wire nails pushed into the furring strip.

Make four string lines — one for each side. Tie loops on the ends of the string lines, and have the lines slightly shorter than the distance between the two nails so when the loops are hooked over the nails the string will stretch slightly and take the sag out of the line.

Start by hooking the string lines on the nails at the lowest level on the guideposts, 2¾" from the floor.

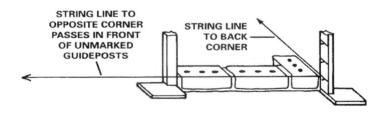

Figure 49

At this point you are ready to lay bricks.

All bricks are not created equal. Some are porous and light, like soft sandstone, and some are heavy as granite and totally impervious to moisture.

The heavy type will need no advance preparation other than a general cleaning, but the porous brick will have to be soaked beforehand. If laid dry, they will suck the moisture from the mortar, leaving a weak bond. Soak them for at least four hours. You can use a garbage can, buckets, or any other container that will permit total immersion.

Have ready a pointed mason's trowel. (See Figure 50) and a $^3/_8''$ finishing trowel. (See Figure 51.)

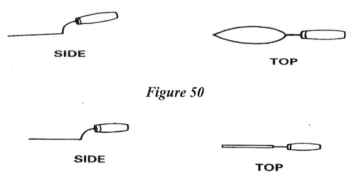

SIDE

TOP

Figure 50

SIDE

TOP

Figure 51

The pointed trowel is for laying down the mortar, and buttering the ends of the bricks, and the finishing trowel is for raking out and finishing the joints.

Also have ready ten pieces of threaded rod. Each piece should be 5″ long. How these rods are used will be explained later.

The recommended mix for mortar is one part mason cement (not Portland cement) to three parts mason sand.

With a garden hoe, mix the dry ingredients in a wheelbarrow or a mortar box. Gradually add water until you have a thick paste. Avoid a sloppy mix. Ideal consistency is slightly on the wet side of crumbly.

With the pointed trowel, scoop up the mortar and lay a bed of it on the floor just inside the string line. If you're new at this, start at the front and only put down enough mortar for two bricks.

Set a brick into the mortar with the end butting up against the front-opening guidepost as shown in Figure 52. Tap the brick into the mortar with the butt of the trowel until it is in alignment with the string line.

Important: Do not have the brick actually touching the string line. Keep it slightly to the inside. A brick touching the string line is likely to throw the rest of the job out of whack.

The end of the next brick and the ones to follow have to be "buttered" before being set in the mortar. What this amounts to is simply taking the trowel and smearing a gob of mortar on one end of a brick. When this "buttered" brick is set against the one next to it, the "butter" fills the space between them.

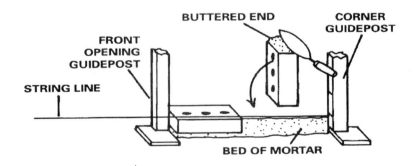

Figure 52

Continue with the job until you have a bedded row of bricks all the way around the base of the stone battery. This first row will have an arrangement like the one shown in Figure 48. Make sure to leave a space for the fan.

Hike the string lines to the next level.

Repeat the brick-laying process — starting once again at the front-opening guidepost. Only this time you'll start with a half brick instead of a whole one. (Later I will give instructions on cutting bricks.)

Between the second and third course, set in place two of the ten pieces of threaded rod (¼" x 5") as shown in Figure 53. Later, these rods will hold the seal between the brickwork and the front of the barrel stove.

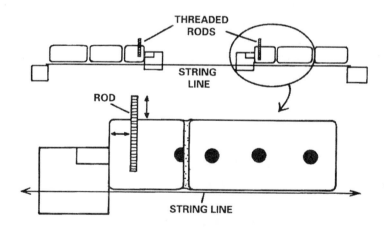

Figure 53

Threaded rods will be set between:
     the 2nd and 3rd course
     the 4th and 5th
     the 7th and 8th
     and the 10th and 11th.

Time out for a few words about finishing the joints. As mentioned earlier, all bricks are not created equal. The composition of the clay varies from brick to brick, and colorwise there is even more variety. Add to this the different surface textures, and you can see that it would be hard to find two bricks that are exactly alike.

If you've gathered your bricks from many different sources, the visual variety of the individual bricks will be even greater — which is exactly what you want. You'll have reds, browns, yellows, and all the different tones in between. Textures will be different, and surprisingly, most of the nicks and chips that come with used bricks will add to the richness of the finished job.

There is only one condition. To effectively bring out the character of the bricks, you have to *isolate* them. It's not enough to plop the bricks in the mortar and forget about them. Something has to be done to the joints to make each individual brick stand out on its own.

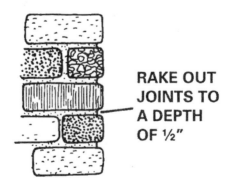

RAKE OUT JOINTS TO A DEPTH OF ½"

Figure 54

What has to be done is a very simple procedure. It's called "raking" the joint, or producing a "weathered" joint. All this amounts to is letting the mortar set up slightly after laying the bricks, and then coming back and scraping the mortar out to a depth of about ½″. Use the point of the finishing tool and scrape or pack the bottom of the indentation square as shown in Figure 54. Also scrape any clinging bits of mortar from the face of the brick. The effect you want is as if each individual brick is protruding about ½″ from a wall of mortar.

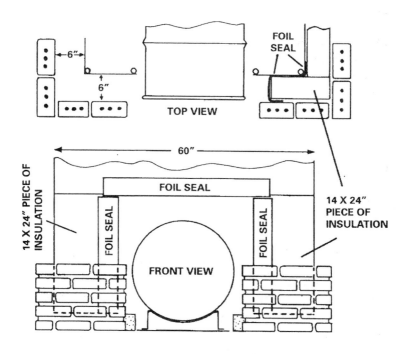

Figure 55

Do this and one other thing after the job dries (explained later) and the end effect of your brick exterior should be no less than eye-stopping.

Stop with the brickwork after laying the 5th course. Finish the joints, clean up your cement tools, and for the time being, turn your attention to insulation.

Between the brickwork and the stone battery is a space of about 6″. Insulation will go here. You're going to cut pieces of 6″ unfaced fiberglass roll insulation to fit between the bricks and the stone battery as shown in Figure 55.

For the front, cut two 14″ x 24″ pieces and fit them as shown. After the brickwork progresses they will support the piece bridging the front opening.

Notice the use of foil seals. Use heavy aluminum foil. Seal where shown and any other spot where there is a break in the sheet metal covering. (Corners, splices, base area, etc.)

Caution: Some brands of insulation contain a resin or binder that gives off an odor when heated to temperatures above 2000°. To say this odor is unpleasant is to say King Kong is just another ape. The odor has a biting, noxious quality that can best be described as "Essence of Grinderdust" or "Smouldering Robot Breath." You won't like it. To make sure you don't get it, test your insulation before installation. Cut off a piece, put it on a sheet of aluminum foil and shove it in an oven preheated to 400°. Leave it in there for 15 minutes. If it remains odorless at the end of that time, consider it safe. As of this writing, Owens Corning (the pink insulation) passed the odor test, but only the Shadow knows if it will pass tomorrow or the day after. Be sure. Test.

Install the insulation horizontally. (See Figure 56.) Reason: the weight of the overriding piece tends to close any gaps that might occur between layers.

Depending on the width of your insulation (12″, 16″, or 24″), go up with several courses of brick, finish the joints, and then drop in another row of insulation.

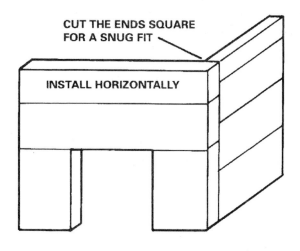

CUT THE ENDS SQUARE
FOR A SNUG FIT

INSTALL HORIZONTALLY

Figure 56

Don't forget the foil seals.

Don't forget to install the threaded rods. (See Figure 58.)

When you finish the 10th course of brick, remove the two front-opening guideposts.

Lay two 1½″ x 28″ angle irons across the top of the front opening. They will be resting on the 10th course of brick.

Also lay the remaining four threaded rods across the top as shown in Figure 57. (See Figure 53 and the accompanying text for more detail on threaded rod placement.)

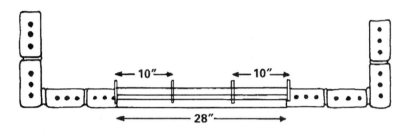

TOP VIEW

Figure 57

When all 10 threaded rods are in place, they will be posi-tioned at the levels shown in Figure 58. (Black dots.)

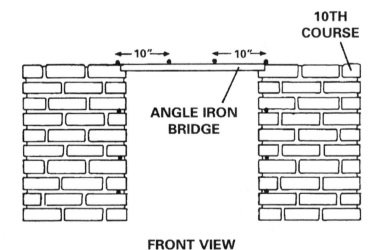

FRONT VIEW

Figure 58

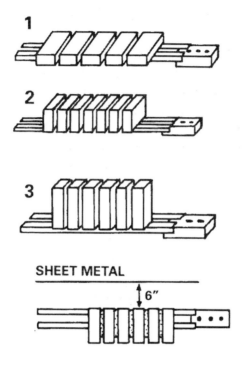

Figure 59

For decorative purposes, you might want to vary the arrangement of the 11th course. The 11th course would correspond to the mantelpiece of a fireplace.

Figure 59 shows three variations in the arrangement of the 11th course.

Some adjustment of the string line would be required for variation #2.

Should you decide to go with a variation, be sure to maintain the 6″ insulating space between the brick and the sheet metal covering the stone battery.

Cutting Bricks

Bricks are best cut with what is known as a "brickset." It's simply a wide mason chisel that looks like Figure 60.

To make a cut, place the brick on a flat, solid surface. Using the brickset and a hammer, tap around the line to be cut. When you have a shallow groove on all four sides of the brick, lay it flat, place the point of the brickset in the top groove and give it a sharp rap with a hammer. The brick should break cleanly on the grooved line.

3″

Figure 60

SHALLOW GROOVE

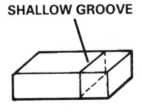

Figure 61

Continue with the brickwork and the insulation until you finish the 23rd course. At this point the brickwork should be almost level with the top of the stone battery.

Cut the last level of sidewall insulation so that it is level with the top of the stone battery.

To keep track of the temperature of the stone battery during burns, a thermometer should be installed at this time.

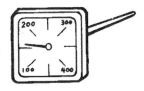

Figure 62

A common kitchen deep-fry thermometer is ideal for this job. It registers up to 400° and looks like Figure 62.

THERMOMETER STEM
RESTING ON TOP OF
SHEET METAL COVERING
STONE BATTERY

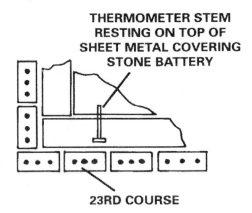

23RD COURSE

Figure 63

Place it near a corner as shown in Figure 63.

(Leave a space in the 24th course of brick so that the face of the thermometer is visible from the outside.)

Cut and fit two layers of 6″ fiberglass insulation on top of the sheet metal covering the stone battery as shown in Figure 64. See that all joints overlap.

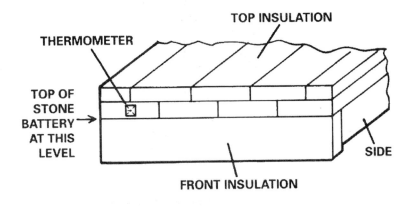

Figure 64

At least once a year, just before the first fire of the heating season, it's a good idea to hose down the stone battery. It only takes a few minutes, and it washes out any dust that may have accumulated during last season's operation.

To best accomplish this hose job, you'll need to be able to lift a part of the sheet metal covering the stone battery. You'll only have to lift a small section of one end about an inch or so — only enough to let you insert a hose nozzle so you can give the stone battery a general cleansing. The idea is to shoot a few gallons in across the top, allowing the water to trickle down through the stones and pick up the dust as it drips to the floor. From there it's only a matter of mopping up around the edges.

To make the hose job something that can be simply and easily done, build a notch into the last four courses of brick. Build it in the middle of one of the walls and make it two bricks wide, as shown in Figure 65.

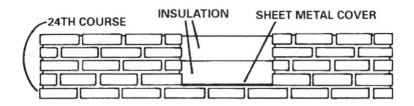

Figure 65

You can see that by reaching through the notch and the insulation, and by prying up the sheet metal cover with a screwdriver, the stone battery is easily reached by a spray from a hose nozzle.

Important: Never spray while the rocks are hot. Not only could you get burned by the water suddenly flashing into steam, but quick cooling could cause the stones to explode. Also, never spray without lighting a fire immediately afterwards to dry out the system. Water left standing on a stove for a day or two can take more out of its life span than a whole season of burning.

To keep unwanted matter from settling on the top layer of insulation, the furnace should be fitted with a dust cover. An effective cover can be made by laying a 68″ length of ½″ pipe across the top middle, and have it supporting two pieces of sheet metal 35″ x 68″ as shown in Figure 66.

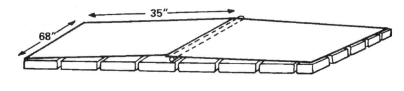

Figure 66

Two more things need to be done to the brickwork before it can be considered complete.

1. *Acid wash*:

It's almost impossible to do brickwork without getting an occasional mortar stain on the bricks. These stains, when dry, show up as white or gray dabs and streaks on the face of the brick. In fact, used bricks are likely to come pre-stained, so even if you personally didn't smudge them up, expect to acquire at least some smudges through inheritance.

Regardless of how you got these blemishes — by accident, inheritance, or even a dirty toilet seat — the standard way to remove them is with a weak solution of muriatic acid. Muriatic acid is sold in hardware stores and home building centers. Diluted (according to directions given on the label) and brushed on the bricks with a paintbrush, it will bubble like spilled beer and dissolve the mortar stains.

Use rubber gloves and eye protection when working with acid.

Clean up with plain water.

2. *Color Enhancement*:

Wet bricks have a more pronounced color than dry bricks. You've probably noticed that, when you soaked the bricks prior to use, the color was greatly enhanced by water.

As the finished brickwork dries, however, all the old pale-
ness returns. Chameleon-like, the bricks gradually lose their
bright earthtones and blend into one another, presenting a look
of obscurity.

To bring back the original rich color, apply a coat of varnish
or polyurethane. When you are sure the brickwork is dry,
brush on one coat of non-gloss varnish or polyurethane. (Gloss
might make the bricks look too "plastic.") That's it. One coat.
The varnish will draw out the colors, permanently, and even if
the end effect falls a little short of genuine art, most tastes
would agree that your finished brickwork certainly is decorative.

The Front Opening Seal

Between the front of the stove and the brickwork is an open
space that has to be sealed. It should be sealed in such a way
that the seal can be quickly and easily taken apart whenever
the stove needs to be pulled out for inspection or replacement.

The ten ¼″ x 5″ threaded rods mortared in place (Figures
53, 57, and 58) are part of this seal. They will serve as the an-
chors.

For the rest of the seal you'll need:

–one 3″ x 30″ piece of 20-gauge sheet metal.

–two 3″ x 24″ pieces of 20-gauge sheet metal.

–one 26″ x 26″ piece of 26-gauge sheet metal.

Fit the 3″ x 30″ and the 3″ x 24″ pieces as shown in Figure
67. The long piece goes across the top. You'll have to mark
and drill four holes in each piece so that they fit over the
threaded rods and protrude ¾″ into the front opening.

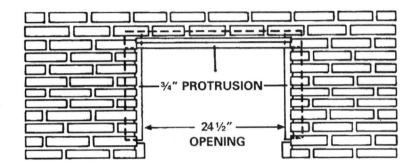

Figure 67

Look at Figure 68. This is a top view. Note how ¼″ nuts on the threaded rod allow you to adjust the position of the sheet metal. You can move the sheet metal *towards* the bricks or *away* from the bricks simply by threading the nuts in or out.

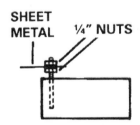

Figure 68

Next you want to use the nuts to adjust the three pieces of sheet metal so that they are *just behind* the lip of the stove. (See Figure 69.) This is so the next and last part of the seal, the 26″ x 26″ sheet, will fit behind the stove lip. (See Figure 70.)

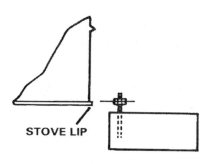

Figure 69

The three (two 3" x 24" and one 3" x 30") pieces of sheet metal you just bolted up are the stationary parts of the seal. That is, there should still be enough top and side clearance so that the barrel can be pulled out without removing these three pieces.

The 26" x 26" sheet is the only removable piece. It completes the seal by fitting around and behind the front lip of the barrel and fastening to the edges of the three stationary pieces of sheet metal as shown in Figure 70.

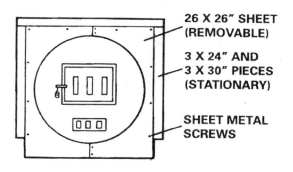

Figure 70

Of course you will have to make a circular cut-out in this larger sheet to allow the front of the barrel to poke through.
To do this:

1. Measure from the floor to the lip of the barrel stove. Say this measurement is 2″. (See Figure 71.)

Figure 71

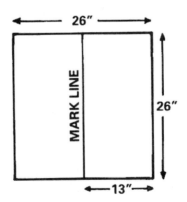

Figure 72

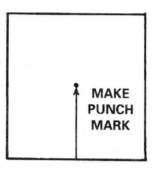

Figure 73

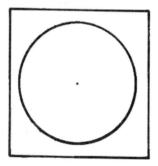

Figure 74

2. Add this 2″ to *half* the diameter of the front of the barrel stove. (Since the diameter of a standard barrel is 23″, let's use the measurement 11½″.)

3. Lay the front sheet out flat on the floor and mark a line through the center from top to bottom. (See Figure 72.)

4. Along the centerline, measure upwards 13½″ (or the number of inches calculated in step 2) and make a punch mark. (See Figure 73.)

5. With a large divider or compass, mark a circle around the punch mark. Make the circle ³/₈″ *smaller* than the diameter of the front of the barrel. (See Figure 74.)

On a standard barrel, the circle will measure 22⁵/₈″.

Note: the reason the circle is made ³/₈″ smaller than the diameter of the front of the barrel is to ensure that the sheet metal fits tight behind the lip of the barrel — the lip being about ³/₈″ larger than the diameter directly behind it.

But don't take these measurements for granted. Check the diameter and lip size of your barrel and make any necessary adjustments.

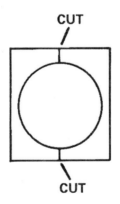

Figure 75

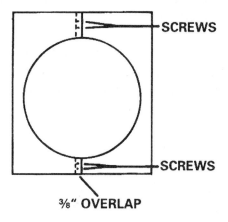

Figure 76

6. Cut out the circle with tin snips, or a hammer and chisel.

7. Now, cut through the front piece at the centerline made in step 4. (See Figure 75.) You now have two pieces.

8. Set the two pieces in place behind the lip of the barrel. These two pieces will have about $^3/_8''$ overlap where they join in the middle.

9. Drill through the pieces where they overlap and join them with small sheet metal screws. Owing to the flimsiness of the sheet metal, you might have to mark the positions of the pieces and drill them on the floor. (See Figure 76.)

10. Finally, drill and fasten the outside edges as shown in Figure 70.

When all parts of the seal are in place and fitting to your satisfaction, you'll notice that there is an open space between the brickwork and the outside edge of the seal. This space was created when you adjusted the seal to fit behind the lip of the barrel. (See Figure 77.)

Plug this open space with mortar. Make up a small batch of mortar — make it stiff enough so that it won't fall or run out — and pack the space all around. Be careful not to block up the removable part of the seal.

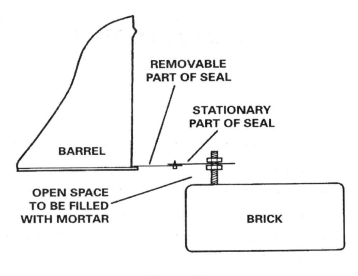

Figure 77

Now that the front of the furnace is finished, let's take a look at the back.

Protruding through the brickwork is the extension pipe.

Using 6″ stack pipe — 24-gauge or heavier — connect the extension pipe to the chimney. Go the shortest, straightest possible route and try to do it with no more than two elbows. Make sure the chimney is rated for wood fires, and make sure no other heating appliances are vented to this chimney. (See Figure 78.)

Stack pipe should slope upward at least ¼″ for every linear foot, and the *male* end of the pipe should be pointing towards the stove.

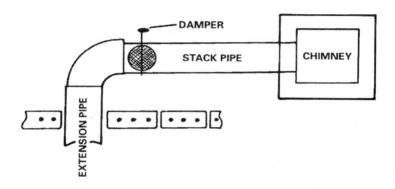

Figure 78

Install a damper close to but not in the extension pipe. (See Figure 78.)

Firm up the stack pipe with support hangers at four foot intervals. Common pipe strap makes good hangers.

All joints in the stack should be secured with three sheet metal screws. (See Figure 79.) This simple little recommendation can't be stressed enough. Three sheet metal screws. Each joint. There's a special section in the I-told-you-so file for those who choose to ignore this recommendation. Some of them are entered posthumously.

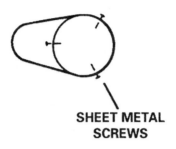

**SHEET METAL
SCREWS**

Figure 79

Figure 80 shows how to wire the system for automatic heat.

Before doing any electrical work, consult local electrical codes. Also, be sure that the fuses are pulled or the circuit breakers are tripped off on any circuits you choose to tap into.

For material you'll need some electrical wire (consult local electrical code as to size and type), a few wire nuts, a fan cut-off switch, and a 110-volt thermostat.

110-volt thermostats are available in hardware or home building centers, usually for under ten dollars.

Hang the thermostat chest high on a wall in a central location away from drafts.

The purpose of the fan cut-off switch is to automatically shut down the fan when the stone battery runs out of heat. No sense in having the fan run needlessly if you happen not to be around when the battery runs out of BTUs.

The best place (once again) to find a fan cut-off switch is ye old scrap yard. Scrutinize the junked furnaces and space heaters. Find a switch box with a *bulb and tube attached.* You'll need one with a bulb and tube because you'll want to feed the bulb down through the insulation covering the stone battery until it lies on top of the sheet metal directly above the stones. (See Figure 80.)

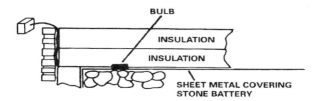

Figure 80

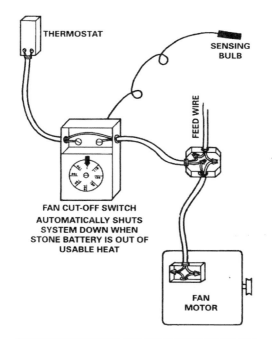

CONNECT WHITE WIRES AND BLACK WIRES AS SHOWN

Figure 81

This completes the building of the Sunstone Superstove. The unit should now be operational.

Chapter Three
Operation

1. Set thermostat for desired room temperature.

2. Set fan cut-off switch at room temperature or above.

3. Start a fire.

A few words here about the fire: Basically the Sunstone system operates like any other woodburner. You burn wood and it heats your house.

The difference lies in the maintenance of the fire temperature. With an ordinary woodburner you regulate the intensity of the fire according to the heat demand of your house. You regulate by adjusting the draft and damper and by selective use of wood and occasional use of prayer.

But with the Sunstone system, you don't diddle with the draft and damper. The draft and damper are both left open and the fire temperature is brought up to a level where it achieves maximum combustion efficiency. Then this temperature is held until the stone battery takes on a full charge.

This doesn't mean you should fill the firebox chock full of flames and have the barrel glowing like a giant stoplight. Not quite. For safety's sake, and to prevent undue stress on the barrel and stack, it's best to keep the stove temperature just below red heat.

And how do you know when it's just below red heat?

A good starting point is to attach a stack thermometer (they're cheap, and they attach magnetically) to the extension pipe where it emerges from the rear of the brickwork. Light a fire and gradually add wood to bring the stack thermometer up to read about 300-400°. Keep the draft and damper open at this point. A fire thus burning is doing so with almost 100% combustion efficiency. Nearly all the "smoke" is being burned and converted to heat.

Try to hold the 300-400° stack temperature by the addition of more wood until the deep-fry thermometer you installed at the top of the stone battery (Figure 62) is brought up to read about 280°.

At the 280° mark let the fire burn down to coals. When the coals are no longer giving off long, yellow flames, all the "smoke" has been burned out of the wood. At this point close the damper and restrict the draft to help contain the heat produced by the coals.

Let the fire go out. Close the draft. This completes the burning part of the cycle.

From here we go to the flameless part of the cycle where the stone battery and the thermostat take over. When the in-house temperature falls below your chosen set point, the thermostat activates the fan and the fan circulates hot air from the stone battery through your living spaces. When the living spaces are warmed a few degrees, the thermostat will stop the fan and hot air will cease to circulate until once again called for by the thermostat.

After a few fires you'll kind of get the feel for the proper burning temperature without consulting the stack thermometer. Also, with usage, you'll acquire a feel for the length of time a fully charged stone battery will provide heat under varying weather conditions. Maybe several days during mild weather, or only 15-16 hours during a January cold snap.

Of course you don't *have* to give the stone battery a full 300 degree charge during the burning cycle. Whatever suits you. But for the sake of comfort and safety, battery temperatures beyond 300° should be avoided.

Maintenance

Keep the area beneath the grate clean. Shovel out the ashes when they accumulate to the point of restricting the flow of air to the fire. Restricted air flow will result in a smoky fire.

At least once a year remove the front end seal and pull out the stove for inspection. Look for thin spots, cracks and excessive rust. Odd as it may seem, summers are sometimes harder on stoves than winters. A stove spending a summer squatting in a damp basement can develop more age spots than a stove operating all winter.

Before the first fire of the season, hose down the stone battery as explained previously.

Check the stacks. Are they developing pinholes? Old stacks, like old people, show the first signs of wear and tear at the joints. If the joints are beginning to rust or crumble, putting off the date of their replacement could mean moving up the date on your tombstone.

Keep the stacks and chimney clean. Over 40,000 chimney fires are reported annually. Who knows how many others are unreported? The point is dirty stacks and chimneys account for a very lopsided majority of damaging fires. Since this part of a woodburning system produces the most trouble, it should be understood what exactly is going on here that's causing all the grief. In a word, it's creosote. Creosote is condensed smoke. When smoke is *not burned* and cools to about 250°, the tar and other combustible vapors condense into a liquid and coat the inside of a flue. Like any other tar-based substance, creosote is highly flammable. Get a good buildup of that stuff on a flue and have it ignite and you will find yourself in a situation that defies control. The stacks suddenly turn bright red — the outside surfaces actually spark from the intense heat — and the pipes themselves shake and roar as if a train were passing through the house.

If that isn't bad enough, creosote bubbles and swells when it burns, and the resulting foam plugs the flue. Smoke that would normally be vented outside backs up into the house and soon brings the visibility down to a choking zero.

This can go on for many minutes, if left to itself, and if the stack sections are not firmly fastened at the joints with sheet metal screws, they will shake apart and fall to the floor, at which point your spectacular chimney fire has escalated to a disastrous house fire. Creosote.

To take some more of the mystery out of why this happens over 40,000 times a year, let's elaborate a little on the mechanics of creosote formation and ignition.

In the fall, during the milder weather, most woodburners are operated at low heat. This means the draft and damper are nearly closed, producing a smoky fire. Some of this smoke condenses before it leaves the system and forms a coating in the flue. There's your creosote. As the heating season progresses, so does the accumulation of creosote. Silently. Out of sight. No alarms are triggered as it forms, and no idiot lights light up to warn you of a dangerous situation. But with each droplet of condensed smoke the creosote thickens, and as it does, it takes on the characteristics of common road tar. It is hard and will chip when cold, and it will turn to a runny, volatile liquid when hot.

Chimney fires mostly happen during the night or during cold spells. Here's why. During these times much hotter fires are required to keep the living spaces warm. With the hotter stove and flue temperature any solidified creosote clinging to the flue will melt and begin to run. If there's enough of it — that is, if enough smoky "cold" fires have been contributing beforehand to the buildup — the creosote will drip down until it falls to the flames coming up from below. The flames will ignite the drippings and within minutes a full-fledged chimney fire will be in progress.

A properly operated Sunstone system is very stingy in its production of creosote, but with neglect it can rank right up there with the worst of them. Burn "cold" fires, operate with the grates plugged up, or regularly burn green wood, and chances are good that your chimney will be included in the over 40,000 that catch fire annually.

Even if you burn sensible fires with seasoned wood, there's going to be *some* creosote formation. It's unavoidable. Outside chimneys are more susceptible because they're exposed to cold outside air, which has a tendency to cool the smoke to the point of condensation, but inside chimneys are not exempt.

Regardless of your chimney location, make a point to check it regularly. Clean it when creosote builds up to a thickness of $\frac{1}{8}$ inch. Cleaning a chimney is not a pleasant job — the chimney is sometimes hard to reach, it's usually dirty and it usually needs its cleaning when the weather is less than balmy — but ignoring it has even greater consequences. Remember those 40,000 people who ignored it last year? Remember some of them with flowers.

Alternative Designs

There's a law for almost everything. There's a speeding law, a law of gravity, one for jaywalking, relativity — there's even a law that forbids a parachutist from ending his fall head first. (Called the law of the "land.")

But there's no law that says you have to build a Sunstone system with a barrel stove or with exactly two cubic yards of stones.

Maybe you already have a moderate to large capacity woodburner that you'd like to use for the firebox. Fine. Just make sure it's close to the floor, for one thing. Cut the legs down if you have to. Make it squat about two inches off the floor. Stones below the firebox *will not accept a charge.* Only those close to the sides or above will take on heat. If you could put your hand on the floor beneath a stone battery charged to 300°, the floor would only be warm to the touch. Heat in stones will always flow upward. The floor or any stones below the firebox will take on only an insignificant amount of heat.

In planning an alternative design to the two cubic yard unit featured in this book, remember that every cubic *foot* (27 cubic feet to a cubic yard) of granite in the stone battery will store 10,000 BTUs.

Maybe you'd like to build a small unit in a rec room and would only want enough storage capacity — say two or three cubic feet — to hold a steady temperature without fussing with the fire.

Or, at the other extreme, maybe you're a landowner cursed/blessed with an abundance of fieldstones that you'd like to put to work. You've got some extra basement space so you surround your firebox with ten cubic yards of stones and only burn fires on weekends — the rest of the time tapping stored heat.

Keep these points in mind when working with alternative designs:

1. Use a stove with a grate and a low profile.
2. Figure 10,000 BTUs for every cubic foot of granite.
3. Make sure to use enough insulation.

Inadequate insulation will limit the effectiveness of your unit. Heat leaking through the sidewalls or the top will result in a loss of temperature control. For instance, if you were to charge any thinly insulated unit to 300° on a mild day, the stone battery might leak more heat than the house requires. This is called "the sauna effect," or "Shrivel City Syndrome."

To keep heat leakage down to a level where you won't have to call your friends over for a sauna party every time you charge the stone battery, stay close to this general rule on insulation:

1. Calculate the square footage of the sidewall area of your projected stone battery — the front, back, and both sides.

For example, the two cubic yard unit featured here has a sidewall area of 80 square feet (four walls each 4′ by 5′).

So 4′ x 5′ x 4′ = 80 square feet.

2. Take this total square footage and divide it by 5. (On the 2 cubic yard unit the total square footage is 80.)

So 80 ÷ 5 = 16.

3. The number you get is the minimum R-factor of the insulation needed for the sidewalls. (On the 2 cubic yard unit we used 6-inch insulation, which had an R-factor of 19.)

4. For the top, double the thickness used for the sides.

Here's another important rule on design. Whenever possible, orient your stone battery vertically. A vertical arrangement will take on a charge faster, and the heat will be more evenly distributed in the stones. A horizontal layout will always have "cold spots" at the outside lower edges.

 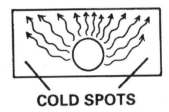

COLD SPOTS

Figure 82

Also, tests have shown that heat in stones always stratifies upward, which means the heat wants to accumulate in a layer towards the top. Think of it like this. Instead of thinking of your stove as radiating *heat* in all directions, think of it as radiating *helium balloons*. It shoots out the balloons in all directions, but where do the balloons go? Right. They go to the top. Now, if the stove keeps shooting out helium balloons, soon the battery space will be full. This corresponds to a fully charged stone battery.

But when you start taking balloons out from the top, as you would when tapping heat from the stone battery, what happens? The balloons that are left inside move up and leave an empty space beneath them. So in effect, you have an empty layer and a full layer. Think of the empty layer as "cold" and the full layer as "hot." There's your stratification. Hot layer on top, cool layer on the bottom, with the hot layer staying hot and on top, but shrinking from the bottom end as heat is drawn off. This is why, no matter how you design your system, *you always have to have a cold air inlet at the bottom of the stone battery, and you always have to have the hot air outlet at the top.*

Conclusion

The Sunstone Superstove takes its name from the fact that the stone battery is heated indirectly by the sun. Energywise, wood is actually stored solar heat. A tree takes its energy from the sun, stores it in wood, and as the wood is burned, it passes the stored energy to the stones, where it sits like an investment — a wealth of heat ready to be drawn off on an as-needed basis.

As investments go, almost twenty years of field testing has been invested in this system. It's simple, and it works. Your average handyman can have one up and running in a few weeks. If you burn wood, keep in mind the flameless part of the heating cycle, as explained earlier. Don't leave home without it.

YOU WILL ALSO WANT TO READ:

☐ **14176 HOW TO DEVELOP A LOW-COST FAMILY FOOD-STORAGE SYSTEM** *by Anita Evangelista.* If you're weary of spending a large percentage of your income on your family's food needs, then you should follow this amazing book's numerous tips on food-storage techniques. Slash your food bill by over fifty percent, and increase your self-sufficiency at the same time through alternative ways of obtaining, processing and storing foodstuffs. Includes methods of freezing, canning, smoking, jerking, salting, pickling, krauting, drying, brandying and many other food preservation procedures. *1995, 5½ x 8½, 120 pp, illustrated, indexed, soft cover.* $10.00.

☐ **14187 HOW TO LIVE WITHOUT ELECTRICITY —AND LIKE IT,** *by Anita Evangelista.* There's no need to remain dependent on commercial electrical systems for your home's comforts and security. This book describes many alternative methods that can help you become more self-reliant and free from the utility companies. Learn how to light, heat and cool your home, obtain and store water, cook and refrigerate food, and fulfill many other household needs without paying the power company! This book contains photographs, illustrations, and mail-order listings to make your transition to independence a snap! *1997, 5½ x 8½, 168 pp, illustrated, soft cover.* $13.95.

☐ **14177 COMMUNITY TECHNOLOGY,** *by Karl Hess, with an Introduction by Carol Moore.* In the 1970s, the late Karl Hess participated in a five-year social experiment in Washington D.C.'s Adams-Morgan neighborhood. Hess and several thousand others labored to make their neighborhood as self-sufficient as possible, turning to such innovative techniques as raising fish in basements, growing crops on rooftops and in vacant lots, installing self-contained bacteriological toilets, and planning a methanol plant to convert garbage to fuel. There was a newsletter and weekly community meetings, giving Hess and others a taste of participatory government that changed their lives forever. *1979, 5½ x 8½, 120 pp, soft cover. $9.95.*

☐ **14192 HOUSES TO GO, How to Buy a Good Home Cheap,** *by Robert L. Williams.* Now you can own that dream home that you've always yearned for — and at an affordable price! How? By following this book's tried-and-true method of purchasing a perfectly livable house that is destined for demolition, and carefully moving it to a suitable parcel of land — all for a fraction of the amount such a home would normally cost! The author has done so several times, and shares his copious knowledge. Follow the process from selecting the proper house, through choosing a mover, to revamping the resettled house. Lots of photographs, and many solid tips on how to go about owning a comfortable home inexpensively. *1997, 8½ x 11, 152 pp, illustrated, soft cover. $18.95.*

☐ **14185 HOW TO BUILD YOUR OWN LOG HOME, FOR LESS THAN $15,000,** *by Robert L. Williams.* When Robert L. Williams' North Carolina home was destroyed by a tornado, he and his family taught themselves how to construct a log home, even though they were unfamiliar with chain-saw construction techniques. In this practical, money-saving book, he clearly explains every step of the process. By following Williams' simple procedures, you can save tens, even hundreds of thousands of dollars, while building the rustic house you've always dreamed of owning! Profusely illustrated with diagrams and over 100 photographs, this is the best log-home construction book ever written. *1996, 8½ x 11, 224 pp, illustrated, soft cover. $19.95.*

☐ **14205 TRAVEL-TRAILER HOMESTEADING UN-DER $5,000, Revised and Expanded Second Edition,** *by Brian Kelling. Travel-Trailer Homesteading Under $5,000* explains how a modest investment can enable you to put an inexpensive roof over your head and live a more independent and self-sufficient life. By following the author's many informative tips, you will be able to purchase a suitable piece of land, acquire a travel trailer or mobile home that can be used as a serviceable shelter, and make the necessary improvements to your property that will enable you to live in comfort and style. *1999, 5½ x 8½, 102 pp, illustrated, photographs, soft cover.* **$10.00.**

☐ **14193 BACKYARD MEAT PRODUCTION,** *by Anita Evangelista.* If you're tired of paying ever-soaring meat prices, and worried about unhealthy food additives and shoddy butchering techniques, then you should start raising small meat-producing animals at home! You needn't live in the country, as most urban areas allow for this practice. This book clearly explains how to raise rabbits, chickens, quail, ducks, and mini-goats and -pigs for their meat and byproducts, which can not only be consumed but can also be sold or bartered to specialized markets. Improve your diet while saving money and becoming more self-sufficient! *1997, 5½ x 8½, 136 pp, illustrated, soft cover.* **$14.95.**

We offer the very finest in controversial and unusual books — A complete catalog is sent **FREE** *with every book order. If you would like to order the catalog separately, please see our ad on the next page.*